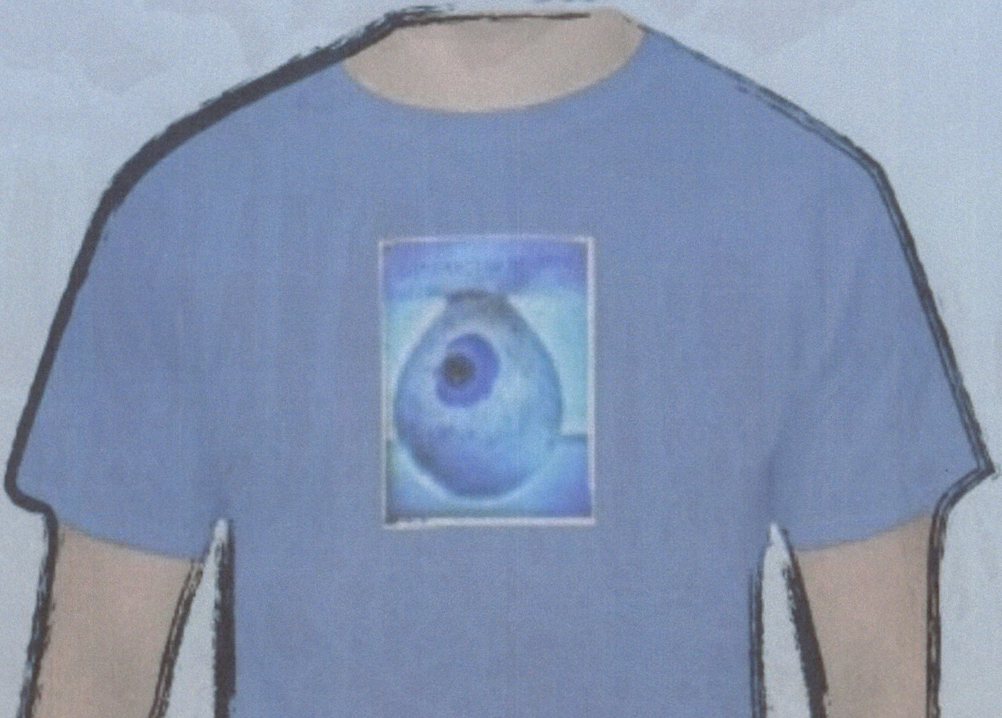

zazzle.com/distractedmasses

Vibrant Protection

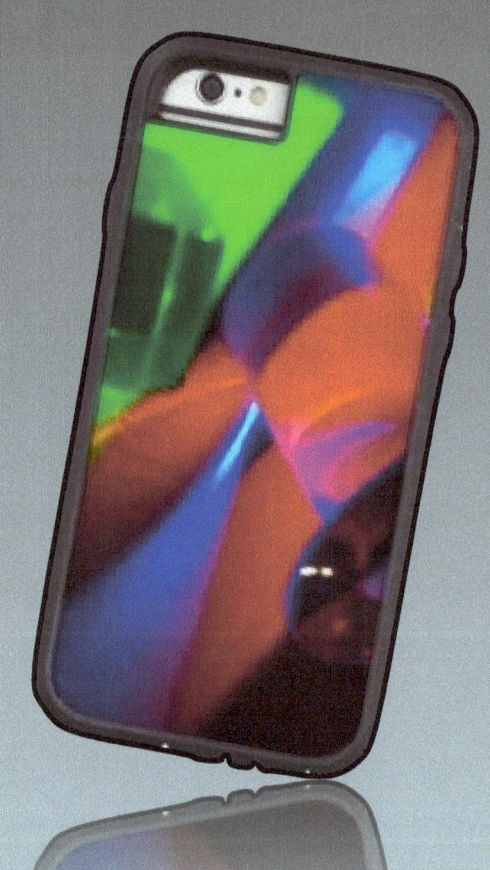

iPhone6 Tough Case
zazzle.com/distractedmasses

What's Inside:

Feature Article:

Memory And How It Contributes To Who We Are

VICTIMIZED: The creator of Distracted Masses becomes a victim of police brutality, cover up, and corruption. Read the whole story here.

BOOK REVIEWS: The Improbability Principle & This Machine Kills Secrets

PLUS: Poetry, Commentary & Cool Merchandise

Only $22.95

ANON DR3SSUP T1E

Summer 2015　　　　　　　　　**Vol. 1 Issue 3**

DISTRACTED MASSES

Time Repeats Itself: Cycles of Life

From the vibrational energy of fornication the human life is conceived. A tiny sperm dances around an egg within the sacred temple of the mother's body. A swimming dance of DNA intermixes, copies, grows, replicates, vibrates outwards, and like a potter's creation, begins to take form. A physicality is molded from a simple vibrational explosion of orgasm. Something from nothing. And it grows with the tummy until the energy form is coated with skin and ready to enter the outer physical world, where light reflects as much as it shines.

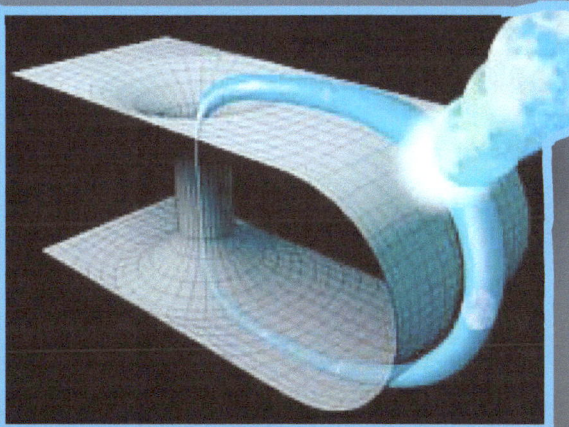

The parents grow old and wrinkle with time, as the generations before them once did. At the same time the baby turns to toddler, turns to teen, turns to full grown. Life continues to bloom. Season to season, sunrise to sundown, new moon to full moon - the days move along a forward-like track of reality in our space-time continuum. Old and weak become our cells. Our vibrational energy leaks loose, and withered and weathered we become. Preservation keeps some fresh, keeps the earth spinning, recycling upon itself. Preservation of the present state - a second of timelessness held dear - but the atomic decay never stops. Into entropy we go. Scattered across all the possible states, but we're told only one future state awaits. That is death, the ultimate decay. No, preservation is not immortality.

For the parents, now grandparents, decompose into the ground, to become one with the earth. Slowly, the energy leaks back to mother nature, to be recycled and regrown. Maximum entropy turns to a state of nothingness - a state of all possible states.

But the light shines back through again from new generational vibrations. Replicating, growing, changing, becoming that shining light of fire within. A new form from the past - a new spin, a new direction. The cycle seems to never stop and yet we live only in this present state. This state of being where directions are shifting. And everything seems never ending, and no one is quite sure of where it is that we call the beginning.

Big bang! Vibrational rhythm, frequency. Which wave function does your consciousness collapse? We are the observers and the creators. Time is endless and limitless too. Repetition is the change. The cycle is the shining light. And it lives within us all. So let us dance back to that frequency we call home. Let's live there and make peace there. Let's stay there until the king falls from his throne. Until the light is regrown . . .

Welcome to Issue 3 of Distracted Masses. The theme here is memory - how it effects us and how it can be collectively altered to re-write history. Read on to learn why everything you think is real may not be so real after all. Enjoy!

MEMORY AND HOW IT CONTRIBUTES TO WHO WE ARE

"in every head there is a world of it's own."

By LaWanda Albright

Each of us comes into the world with unique and universal qualities. Our brains are wired to expect certain stimuli from the environment - someone to care for us, someone to help us learn language and communicate, someone to help us negotiate the social world. The neural pathways that develop depend largely on the culture and time into which we are born. The language(s) we speak, our social interactions, and where we focus attention are prime examples of the body-brain interaction. The brain is expecting to learn these things and looks for cues in the environment. People in our environment - parents, siblings, neighbors - act on the shared environment and influence our development. Our brains receive input from our bodies and integrate all of the different signals (sight, hearing, smell, etc.) into coherent ideas and constructs the world from our best guesses, interpretations, and expectations.

We begin forming memories even before we are delivered into the world. The environment of the womb begins to shape who we are. Newborns have been shown to recognize their mother's voices. What mom eats while pregnant influences food preferences latter. All future development depends on the interaction between our biological temperament, the environment, and the relationships we develop. The neural pathways along which each of us develops are unique to us. As we begin to manipulate our world and notice reactions, we start to influence the growth of our neural pathways. We are born with genetic predispositions, but we can change the way our brains develop by manipulating our environment. We begin to construct a reality for ourselves as we choose where to focus our attention and how to engage with the world.

Our first memories are unconscious and are developed slowly, over time. We remember how to grasp an object and bring it to our mouth, building the foundation for more sophisticated manipulations. We watch our caregivers for cues and emotional reactions, forming the basis for future reactions of fear or joy. This social referencing guides the development of the amygdala, the part of our brain devoted largely to emotion and our fight, flight, or freeze reflexes. This part of the brain plays a primary role in learning and motivation. Memories from the first few years of life remain largely unconscious, but are important for future development.

We don't seem to have much ability to recall memories before the ages of 3 or 4 and those are few and often tied to a major life event. Memories get reinforced as we talk about them, so it makes sense that declarative memory begins to develop along language pathways. This is what most people refer to when discussing memory.

"Memory is personal and intertwined with emotion. It is an extension of perception." - **Larry Squire, PhD**

Memories are laid down along multiple neural pathways and consolidated in the hippocampus. For an event, fact, or skill to become embedded in the hippocampus several things must happen. At some level we must attend. Where we direct our attention determines which neurons fire and connect. Once we are attending, we must decide whether to engage, reject, or observe. Although there are universal brain structures with specialized activity, which neurons fire and eventually wire together is unique to each of us and largely dependent on culture and family relationships.

We remember things that are meaningful and challenging, meet a need, or arouse curiosity. These social and emotional interactions cause our blood chemistry to change. Changes in blood chemistry alert neurons throughout the brain to be on the lookout and as more information is obtained, the neurons connect in old and new ways with new neurons joining the network. When a new association is made with an existing pathway, the new is incorporated and the pathway becomes myelinated. The white matter, myelination, is the insulation for the neural pathways. These are the memories that we continue to develop as we use old knowledge to create new understanding. There are multiple memory systems and multiple ways to access and express memories.

Declarative memories are consolidated in and retrieved from the hippocampus. When we recall a memory, we are likely to bring up the most salient points and then fill in the rest as we discuss and think about the event or concept. Pieces we fill in may or may not be the reality. When we recount a recent trip with our best friend, we remember different aspects and may disagree about what actually happened. In the end, those discussions get incorporated into our original memory and each retelling may alter the memory in some way, but the core of the story - the trip with our friend - remains the core memory. The more we recall facts, concepts, and procedures, the stronger the memory becomes. Practice, and reflection strengthen neural pathways and support memory. When we prepare our minds to engage in an activity we are more likely to remember the activity. Memory formation is assisted by predicting the outcome, responding, failing, and getting immediate corrective feedback. When we find new and novel ways to use old knowledge, neural pathways are strengthened and in some ways begin to change our memories.

Cortisol, often referred to as the stress hormone, plays a vital part in memory formation and retrieval. Cortisol is always present in our bodies and is necessary for certain functions within the adrenal system. When we become stressed, we produce additional cortisol. A bit of stress, and cortisol, helps us focus our attention and aids in learning & memory. Too much stress, or stress over an extended period of time, has a negative impact on memory. An over abundance of cortisol can make you hyper vigilant and attentive to specific details and events while completely shutting out other stimuli. It can keep the hippocampus from encoding and retrieving memories and over time can shrink the hippocampus.

Remembering incorporates new information into previous memory and retrieval overwrites the memory. Our realities are constructed from our own interpretations and expectations of the world. We might offload some of our memory into the environment via our mobile devices or visual reminders. Offloading can be used to verify what we think we remember, but it doesn't stop inaccurate associations as the result of partial memories. When

Brian Williams' memory doesn't correspond with the evidence and the shared experiences of others, it may be that the neurological pathways were disrupted and rewired through recall of events in association with new information. Neurologically, a fabricated memory and a real memory are the same. It is the shared and validated memories that give us a sense of self.

Our ability to recall and discuss memories gives us a sense of self and how we relate to others and the world. Without memory there is no future. All new experiences are processed in relation to past experiences. If we can't remember the conversation we just had, how can we think about what we might want to do tomorrow? People with dementia or early Alzheimer's will not remember recent events or be able to tell you where something is, but they can engage in conversation about years old events and show you exactly how to get from point A to point B. What we remember depends on its relevance to our lives and our current realities. Long term memories take longer to fade. What happens when our memories are not supported by our current realities?

"The new perishes before the old." - **Larry Squire**

Memories can be manipulated. The way a question is framed can influence the response. Pictures and videos can be altered, and people can be made to believe they have experienced something they really haven't. Some work with people suffering from post traumatic stress is all about altering those memories so that they don't interfere with current life events.

Supplements and new drugs are being developed and tested to determine their impact on our memories.

Our memories are important to our sense of self. When the internal structures of our mind and the external realities clash and become distorted, we experience distress. When this happens we try to force each other into the roles we assign. This can lead to coercion and violence. Our experiences in the world form who we are and our memories help us interpret our current experiences. We all work to make sense of our lives, and the behavior we exhibit when our memories aren't helping us navigate the now can have an impact at all levels of our society.

"Memory - the ability to use our past experiences to determine our future path." - **Kim Ann Zimmermann**

Glossary

amygdala - the amygdala is an almond shaped part of the brain responsible for the perception of emotions (anger, fear, sadness, etc.) as well as the controlling aggression. The amygdala helps to store memories of events and emotions so that an individual may be able to recognize similar events in the future

conscious memory - also known as declarative memory

declarative memory - memories that can be consciously recalled such as facts and knowledge. Declarative memory can be divided into two categories: *episodic memory*, which stores specific personal experiences, and *semantic memory*, which stores factual information

episodic memory - our memory of experiences and specific events that occur during our lives, from which we can recreate at any given point

hippocampus - a seahorse shaped structure located within the brain and considered an important part of the lambic system. It functions in learning, memory, and emotion.

long term memory

myelin - compact fatty material that surrounds and insulates axons of some neurons, the "wiring" mechanism in the phrase "neurons that fire together, wire together."

neural pathways - A neural pathway connects regions within the brain to one another or conveys information from the peripheral nervous system to the brain. Two major classes of neural pathways relay sensation to the brain or carry signals for movement to the body from it. They both consist of long, insulated nerve fibers that communicate electrically or by a chemical neurotransmitter.

semantic memory - refers to general world knowledge (facts, ideas, meaning and concepts) that we have accumulated throughout our lives.

short term memory - the capacity for holding a small amount of information in mind in an active, readily available state for a short period of time. The duration of short-term memory (when rehearsal or active maintenance is prevented) is believed to be in the order of seconds.

unconscious memory - a type of memory in which previous experiences aid the performance of a task without conscious awareness of these previous experiences.

working memory - the system that is responsible for the transient holding and processing of new and already stored information, an important process for reasoning, comprehension, learning and memory updating. Working memory is generally used synonymously with **short term memory.**

References

Buonomano, Dean, " Brain Bugs: how the brain's architecture shapes learning, memory and decision making." Learning and the Brain Conference. Boston, MA. Nov. 2013

Catterall, James S., Ph.D. "Artistic Expression, Hot Cognition & Student Motivation." Learning and the Brain Conference. Washington, D.C. 7 May 2010.

Klemm, W. R. "Memory in the Age of Google." Learning and the Brain Conference. San Francisco, CA. 13 February 2014

Lafoff, George. "How Brains Think: Embodied Cognition, Language & Metaphors." Learning and the Brain Conference. San Francisco, CA. 14 February 2014

Roidiger, Henry. "Making It Stick."Learning and the Brain Conference. San Francisco, CA. 12 February 2014

Squirem Larry. "Structure and Organization of Memory." Learning and the Brain Conference. San Francisco, CA. 12 February 2014

Srouse, Allen. "Development of the Person ." Learning and the Brain Conference. Washington, D.C. 7 May 2010.

Wiley, John & Sons. The Brain and Learning. Jossey-Bass A. Wiley. San Francisco, CA. 2008.

Wexler, Bruce. *Culture and the Brain*. Cambridge, MA: Massachusetts Institute of Technology 2006. Print.

**Praying Mantis Dreams 13" X 21"
Polyester Lumbar Pillow – $36.95**

**Tractor Trucker Hat
$37.95**

P3ACE! Bumper Sticker – $3.95

**Surfer Basketball
Hoop – $48.95**

**Change Tank-Top
100% Cotton – $23.20**

Meta Narrative

Rewrite the present to shape the past.
Future events deciphered,
Oracles or disinfo?

Control discussion. Frame agenda.
Change the subject.
Quick, make a distraction!

Look for the patterns.
See them across the urban media scape.
Not futurism or determinism.
Meta narrative for the masses.

Fool's Memory

Memory.
Relapse, flashback, big bang,
Past life sensory.
Dream like imagery.
Feel this energy.

Time slice error repeat,
I beat those who seek to chase deceit.
Collective lost thought forms.
Distracted masses is the new norm.

Reality is really only what you perceive it to be.
That includes all of history.
Everything you were told in school is a lie,
Or at least a mistake, mistruth, or misconception of
some sort.
You can try to relearn it for all that's worth.

But some things aren't lies.
They really did happen.
Some things don't go away.
They only grow stronger in the neural pathways.
From reality those memories don't stray.

They can misconfigure,
Deceive, trick, try and persuade & convince.
But the people know the picture's much bigger.

Bob Marley said you can fool some people
sometimes,
But you can't fool all the people all the time.

Feel this energy,
Dream like imagery.
Past life sensory.
Relapse, flashback, big bang.
You can't erase the collective memory.

False Memory

This doesn't exist.

In Remembrance

9/11, Fukushima, James Boyd.
Bob Marley, Sudden Rush, Pink Floyd.
Enter the nexus,
Pandora's big void.
Come complete,
Erratic anniversary.
Annoyed.

Can't deny the data,
It's not Photoshopped.
It's in your own evidence room.
Big boom.

Libraries destroyed,
The truth is on fire.
They want to erase the past.
They want to erase the important things that should
last.

For what purpose,
I'm not sure, but,
Let's take control fast.

They're leading us off a cliff.
Thinking we really are all just sheeple.
Power to the fucking people!

Let's not forget.
Let's not let time slip.
Hold on to what you know.
It's a paradigm shift.
Let's lead them off that cliff.

And then pay respect to the so called demigods.
Monument erect for the heroes who saved them.
But you know no credit goes to the real men.
And we still forgive all their sick sins.

This is in remembrance.
Of all the false histories.
Of all the fake memories.
Disassociate this.

This is in remembrance.
For all those who have fallen.
9/11, Fukushima, James Boyd.
This is in remembrance.
For all those who will fill their void.

Fakery

Selective input,
Pay attention HERE!
Staged reality craft.
Fakery to instill fear.

Remembering & Forgetting

By Rick Albright

The recent revelations concerning Brian Williams and Bill O'Reilly"s self deceit was not a surprise. Upon leaving Vietnam it became apparent to me that some of the stories told by others on the trip home began to give the teller more credit than he claimed the first time the story was told in the barracks. My wife has spent her entire life helping me to keep my stories in check, but then we began to realize that not all of her memories were accurate.

We are brain hobbyists who attend *Learning and the Brain Conferences* on a regular basis to keep up to date with the latest neuroscience research about learning and memory. As former teachers we were always interested in brain growth and development to help us understand how to better present lessons to our students so they would remember what they had been taught. Despite being mindful of trying to teach for recall, former students I have met over the year seldom remember what course I taught, much less specific detail. Some of that may be explained by the fact that I taught history. I can not tell you how often, upon learning I was a history teacher, adults tell me how much they disliked history classes but now that they are older they love reading about the past.

That is an important aspect of learning, you must first be interested if you are to remember. Although that seems obvious, but to understand why that is true one must know that learning and memory are two parts of a whole process. Neuroscientists who investigate these things say that memory is dependent upon more than electrical interactions between sodium and calcium, it is emotional and personal driven by both nature and nurture. Neuroscientist are exploring the cognitive, psychomotor, and affective domains to help them understand how brains remember. Their discoveries provide a wide variety of methods for teachers to enhance student learning, but more importantly these methods can help a person to live a healthier life by developing strategies to strengthen their brain's learning/remembering systems.

The first step to understanding memory is to know how brain cells or neurons transmit information. Neurons are composed of dendrites and a cell body (the soma) where information can be received and a tube shaped axon which connects to the dendrites of another neuron. Communication takes place within the neuron and between neurons. The communication within the neuron is electrical and between neurons chemical.

A sensory stimulus (ouch) sends an electrical impulse to a neuron and that signal travels quickly along the axon of that cell until it reaches the synaptic gap, a tiny opening between cells. This is where the electrical potential triggers a chemical reaction that releasing hormones that cross the gap to the adjoining cell's feathery dendrites. Thus beginning a process that can form thousands of links with other neurons giving a typical brain well over 100 trillion synapses.

The movement continues until the stimulus has created a neural network categorized by function that was created by the stimulus and neurotransmitters. This network is a pathway that shares itself with the hippocampus where the axons in the neural network are strengthened through the process of myelin (white matter or glial cells) building up on the axon. If the axons do not get strengthened appropriately the network fades away from the brain.

Learning is the act of making and strengthening the neurons forming the network and memory is the ability to reconstruct or reactivate the previously made networks. It is important to remember that networks that fire together wire together. When another similar stimulus (ouch) is received the old networks light up saying hey there is more about that over here. And that in turn myelinates the axons.

So what can we do to strengthen our memories to remember what we have done or even who we are? John Medina, a neuroscientist from Harvard, advocates first of all a healthy brain. The brain is one of the most amazing organs in the human body. It controls our central nervous system, keeping us walking, talking, breathing and thinking. The brain is also incredibly complex, comprising around 100 billion neurons which are each connected to a thousand more. Thinking and remembering occur because of the integrated action of these neurons. Every thought and action is controlled by the brain. The brain uses more energy than any other human organ, accounting for up to 20 percent of the body's total use. There are over one hundred thousand miles of vessels and capillaries intertwined with the brain to supply the oxygen needed to fire neurons, for cell maintenance and to produce the proteins needed for synaptic transfer.

All of this convinces me that John Medina is right. He recommends a healthy diet, exercise, adequate sleep and stress reduction as some of the ways to maintain a brain that will remember, even in old age. This is a life long process and commitment that each person has to decide to make in order to live a full and healthy life.

My father contracted congestive heart failure when he was in his late 80s. The lack of blood flow caused his brain to change. He lost his immediate memory and was unable to create new memories. He was able to recall many events and details from his early life, childhood and his young adult life. He remembered chemistry, his major in college and some family events if he was primed. He could, however, talk in detail about his experiences in WWII as a bomber pilot and he could remember some things about his career as an Air Force officer. But, he couldn't recall what we had just discussed minutes before. It pointed out to me the importance of a healthy brain. You have to keep the blood flowing and insure your body can produce the elements needed to keep your nervous system and brain firing.

If you don't sleep well your brain can not consolidate the neurons into networks to give you plenty of points of recall for the things you have recently learned. It is obvious that exercise causes the blood to flow and carry oxygen to the brain. A healthy diet impacts your total health, obesity, for instance, causes changes in the brain that impair memory.

Stress is the most toxic of memory disrupters. It releases cortisol, a chemical that inhibits the information processing system. The impact of not following a lifestyle to have a healthy brain and body may take a while to impact a person, but eventually your memory will exhibit symptoms of poor care and maintenance if you don't.

The doctor who examined my dad after he had experienced rapid deterioration of his memory did not believe Dad had a stroke, although many of the symptoms were similar. They weren't sure what caused it, but in retrospect he was probably already experiencing symptoms of low blood pressure and his brain lacked the blood flow necessary to provide energy his brain needed to work at full capacity.

Dad had always been really smart, Mensa type of smart. He had been a pilot, a nuclear chemist, and an officer in the Air Force among other attainments. The loss of his memory was hard for me as I recalled what a witty and humorous man he had been and now he read the same story in the newspaper all day long. I could sort of prime Dad's memory and get him talking about many things from the past and event recent family or world events. His memory pretty much stopped at about the time he had the rapid deterioration.

He hated exercise, it was kind of a joke in the family. He was physically strong but his cardiovascular system was weak. He had smoked when younger, his primary diet for many years centered on foods that are not heart healthy, he had high blood pressure most of his life (stress), and he had sleep apnea for quite a while. So the amazing fact was that he lived to 93, about five years with impeded memory and immobility.

But what he could remember is a testament to memory being personal and emotional. If you want to remember something find a way to make it yours and feel strongly about it. Dad's most vivid memories were from WWII. He could remember missions, people, places, procedures and policies. He could remember Mom and many things about our family as we moved to where the Air Force sent us. It was astounding the things he could pull out of his mind.

He could also remember things he had learned in school. Especially those subjects he had been curious about and where he studied and prepared for exams. At the last Learning and Brain conference in San Francisco which we attended this year, these two techniques had undergone scrutiny by neuroscientists who reported students remembered best if they were tested repeatedly. Not high stakes testing, but repetition of the information. Quizzing yourself would work. My dad could remember quite a bit about chemistry a subject he loved and one he had been tested on throughout his studies and then in the field.

Current research adds some useful information to the concept of repetition. It demonstrates that it is best to pause between those tests or repetitions. Our brains need time to consolidate information in order to build strong networks of neurons. Everyone has experienced that aha moment when you have taken a break from a difficult task and suddenly the answer is upon you. That is a reason sleep is so important, it is during deep sleep that your brain consolidates previous information. If stimulus events do not get put into networks they fade and eventually are forgotten. Even information you work hard to remember can fade over time. That is why my former students have difficulty remembering my class, they did not keep firing the networks I tried to help them develop.

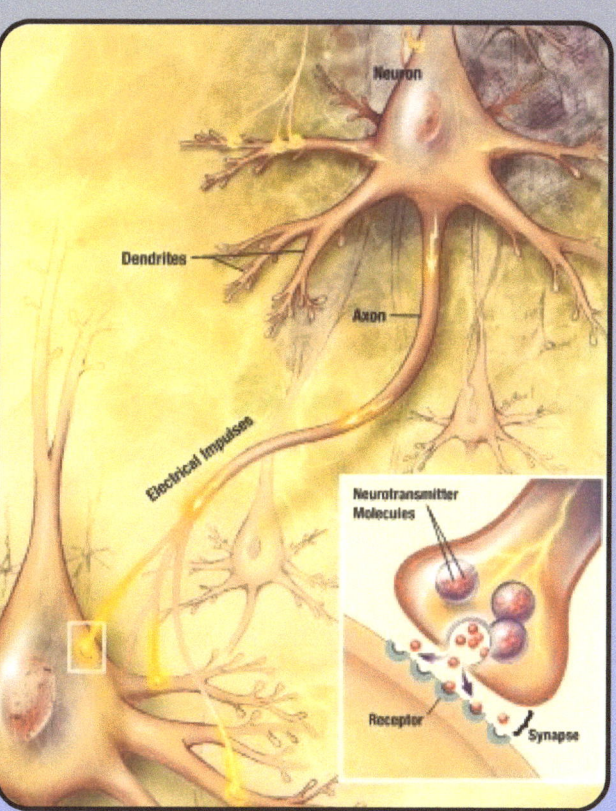

Nothing works to build strong memories if you do not pay attention. You have to attend to the information you are trying to learn. There are a number of ways to enhance attention through the use of media, peer teaching and hands on learning. The value of those methodologies can be negated, however, if the student has a mindset that puts them in opposition to learning or attention can be diminished by multitasking. Using electronics, passing notes, day dreaming and an over stimulative environment can keep information from ever reaching the hippocampus. You might think you can do more than one thing at a time, but it is unusual to do all of these things well. It is better to complete one task before tackling another if you want to be able to remember what you have done.

Even if you are healthy there are many impediments of strong recall. Evolutionary biologist, Robert Trivers from Rutgers University believes that self deception, an evolutionary adaptation, distorts what we remember.

"Our sensory systems are organized to give us a detailed and accurate view of reality," he says, *"but once this information arrives in our brains, it is often distorted and biased to our conscious minds." We repress painful memories, create false ones, rationalize immoral behavior and jack up our self-esteem. We deny ourselves the truth."*

We are influenced by, as Robert Sapolsky, evolutionary biologist from Stanford, posits, a range of nature nurture events shaping who we are and contributing to our mindset beginning with evolutionary impacts from prehistory right up to what we are doing when the stimulus is introduced into our sensory system. What we remember is truly a matter of our genetic makeup, the environment in which we have been raised, and our surroundings at the moment of input.

Since there is no way that the brain can pay conscious attention to all sensory data that constantly bombard the body, it filters out that which is not relevant. That means that, according to Pat Wolfe, a brain educator, approximately 99% of all information entering through senses is immediately dropped. That is probably why so many eye witnesses differ with other eye witnesses when describing what happened to whom.

Add to that the idea that every time you remember an event you actually remember the last time you remembered it event. It is a little like the gossip game where nobody can keep a story straight. That is what happens to all of us. That is why Brian Williams and Bill O'Reilly are defending their integrity, it is a human thing. Just like the young marine I recall returning from Vietnam at the same time I did. In the barracks in Danang his stories were creditable. In Okinawa he had become a central character in his stories, by the time we reached the states he was a hero. I think we do this inadvertently to put ourselves at the center of our stories.

There is a way to ensure you will remember. You could use technology to record your life. Maybe there is a government agency that can help you. If you don't want to do that you will have to live a fully human life coping with your brain's strengths and weaknesses while trying to stay healthy.

Suggested Readings On Memory & History

1. Esther T. Bui, Myra Sourkes, and Richard Wennberg. "Generalized tonic-clonic seizure after a taser shot to the head." *Canadian Medical Association Journal.* 2009 Mar 17; 180(6): 625-626. http://www.ncbi.nlm.nih.gov/pmc/articles/PMC2653575/.
2. Readers Digest. "5 memory Boosting Secrets From a Police Sketch Artist." 2015. http://www.rd.com/health/5-memory-boosting-secrets-from-a-police-sketch-artist/.
3. Savage, Kirk. "History, Memory, and Monuments: An Overview of the Scholarly Literature," University of Pittsburgh. http://www.cr.nps.gov/history/resedu/savage.htm.
4. Green, Anna. "Individual Remembering and 'Collective Memory'. Theoretical Presuppositions and Contemporary Debates," *Oral History,* Vol. 31, No. 2, Memory and Society (Autumn, 2004), pp. 35-44. http://lineberger-center.lr.edu/sites/lineberger-center.lr.edu/files/images/Green-%20Individual%20Remembering%20and%20'Collective%20Memory'.pdf.
5. Millhiser, Ian. "Special report: Inside the Koch-backed history lessons North Carolina wants to teach high school students," *NC Policy Watch,* Dec. 17, 2014. http://www.ncpolicywatch.com/2014/12/17/special-report-inside-the-koch-backed-history-lessons-north-carolina-wants-to-teach-high-school-students/.
6. Corbalis, Michael C. "Mental Time Travel: How the Mind Escapes From the Present," *Cosmology,* 2014, Vol. 18: 139-145. University of Auckland, School of Psychology.
7. Winer, Stan. *Between the Lies; Rise of the Media-Military-Industrial Complex.* London: Southern Universities Press, 2004.
8. Greenwald, Glenn. "How Covert Agents Infiltrate the Internet to Manipulate, Deceive, and Destroy Reputations." *The Intercept,* Feb. 24, 2014.

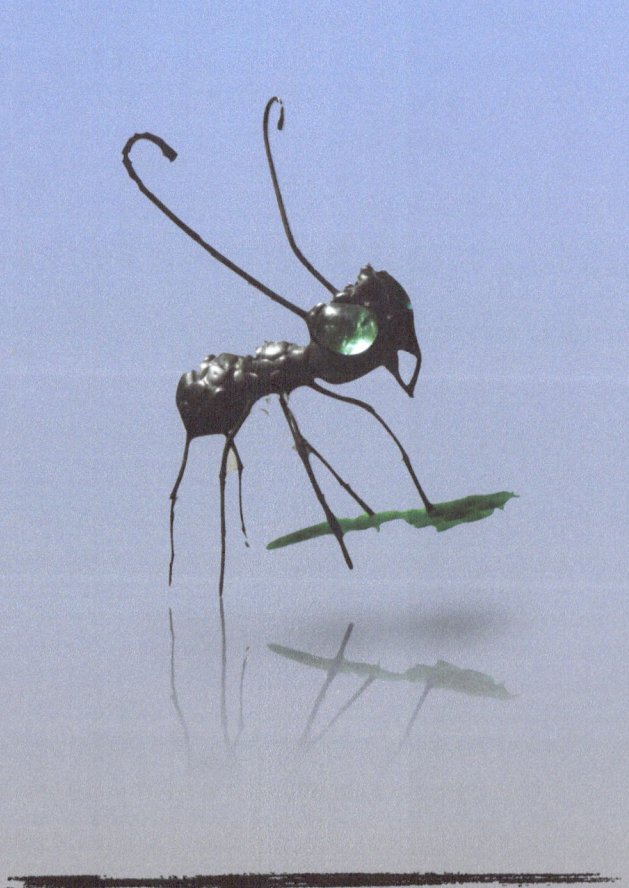

D1STRACTED

1   800   <<<   PROF200.4C   A-F   CMX

MASS3S

MASS3S

D1STRACTED

Snow Streaks

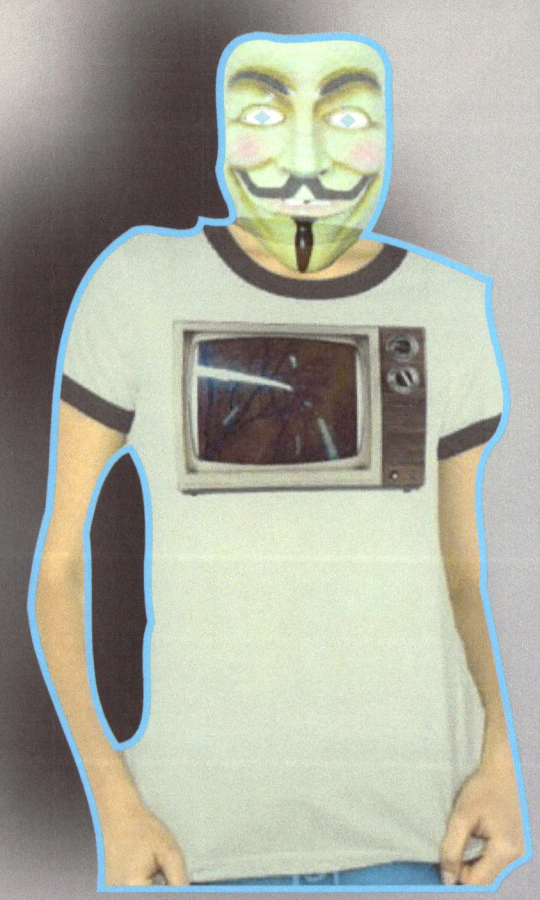

zazzle.com/scajax

DANGER!

They're out there.
The criminals in the shadows.
The shadow government.
The puppet people.

They live, they're real, they come to haunt your day. Your night. Your life.

Not just illusion, delusion, hallucination. They are the one's causing that emotional sensation.

That frequency. That vibration.
Mind-controlled masses in a panic as if it was really all that tragic.
Informational cloned DNA data,
Everywhere I go I'm a persona non grata.

A ticking time bomb.
A smeared public image,
A stigmatization,
An awakening.
A realization.

Throw him to the floor!
Lock him up! Chain him!
Don't matter the numbness,
Just change him!

Rehabilitate to that Christian state.
Must be a terrorist, rebel, insurgent, or some sort.
Sorties'll sort em out, sort of.
Come forth!

I demand to see your face from the shadows!
Out go,
All emotion . . .

Life is but a dust storm lost in the ocean.
It must be extinguished.
Too dangerous for the people. Too dangerous to be treated like an equal.
Stomp em, chomp em. Lock em up and dock them. Mark em as DANGER!

FISTFIGHT

The kids on the playground don't play anymore.
They hunt zombies and kill and shoot guns for the thrill.

They chase robbers and smash their face to the ground.
They meet after school by the swings to throw fists.
They're young and they're pissed.

Trained to love violence.
To believe it's a reward.
Why shouldn't they cause discord?

Really nothing. I should be numb already. Desensitized to the shit.
I shouldn't really care.
But I guess I do.
Don't you?

FREE SPEECH

Free speech,
Live speech,
My speech,
Each week.

People saying nothing but hate speech.

Hate freaks.
Hate sheikhs,
Hate geeks,
Hate the meek,
Hate the weak,
Hate leaks,
Hate speak.

Is that what we get for free speech?
Maybe I should pay more money to hear it, more money to be in fear of.
The monsters you hate.
You monster. You're fake!

Give me real speech,
Hip-hop, freestyle, sick speech,
Spin, style, skid, and scratch a real screech.
Heal speech.
Freedom talkers and people with real reach.
Love, compassion, action, and life everlasting are what they seek.
Always living life at a high peak.

I peek,
Out into never nowhere and see my peeps.
I need less dark futures of people like Nietzsche.
I creep,
Out from the shadows and learn from the light that I'm free.
I see,
Real words of reason and false accusations of treason.
Bending words like reflections in a holograph. And   calling it free speech.
Poetry is the freest way I can get my mind to speak. It's real free speech.

SCATTER

Entropy, distraction.
Mind memory, erasure.
Tracer, Taser . . .

Dot, danger, anger,
Now I'm a stranger.
Hung out in a hangar.

Spread, smeared, scattered, and random.

Helicopter doctor, poet, and suspect operate in tandem.
A big fan of . . .

Entanglement,
Brain scatter.
I forgot.
Don't matter.

MEMORY-JUMP

Deja vu, Did it before.
Do it again,
In a dream of silk sin.

Remember how it was different?
A second ago? A split second before.
What? Is it a time jump?
A multiverse of matter?

Don't forget your spatial placement.
Memory erasing.
Did I do it before?
I'm really not sure.

DISSILLUSION

Last night was yesterday's dream,
A reflection of darkness,
A mirror shattering scream.
A self conscious realization,
A space station, a cryptographic beam.

Memories are all that they are,
Even if they aren't even what happened.
Reality is what they say it is,
Your memory is a falsehood.
Your existence denied.

Records deleted.
Time loops repeated.
A jagged edge on the
mountaintop.
Just new thoughts planted
and seeded.

Your illusion is not true,
So they want you to believe.
Their illusion is the answer,
So you seek to retrieve.

Information unfolding,
Microdot moonshot,
A life that your holding.
Protect them. Protection.
Look at these lives that we're molding.

Fall in or get out! Well fuck that.
I'm not a part of your reality.
Not even a reflection, reversal,
Not even a rehearsal.

This is the real deal.
My disillusion. Don't need your intrusions.
It's messed up enough without your accusing.
Just for your amusing.
This is my reality. The one I'll be choosing.

FORCED REALITY

Collective memory erasing,
History rewriting,
Built around a monument of death.
For mass remembering.

Outlasting generations,
Lies and illusions for all nations.
To hold us prisoner to their creation.

Banksters and fraudsters,
Arms dealers and dynasties.
All for the power,
All to be number one.

Because number two would
be the loser.
Because reality isn't ours
for the making.
It's our souls that their
taking.

Making us watch the shadows in
the cave,
Forced like cattle to our graves.
Herded and branded and cast off into
never nowhere.
Those oligarchs don't care.

Just cannon fodder, pawns, pansies, minions
for their pleasing.
Laughing at your pleading.
Bombs for more bleeding.
They resurrect death and stop all breathing.

Scorched earth, omnicide, we're just bugs to eradicate.
They love to see us kill and hate.
Conquer and divide.
More bloodshed to carve their path to power.
To carve their reality for the masses.
Shoved down the throats until time stops passing.

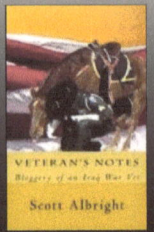

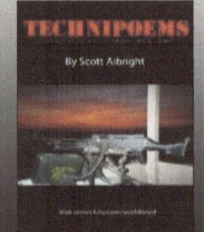

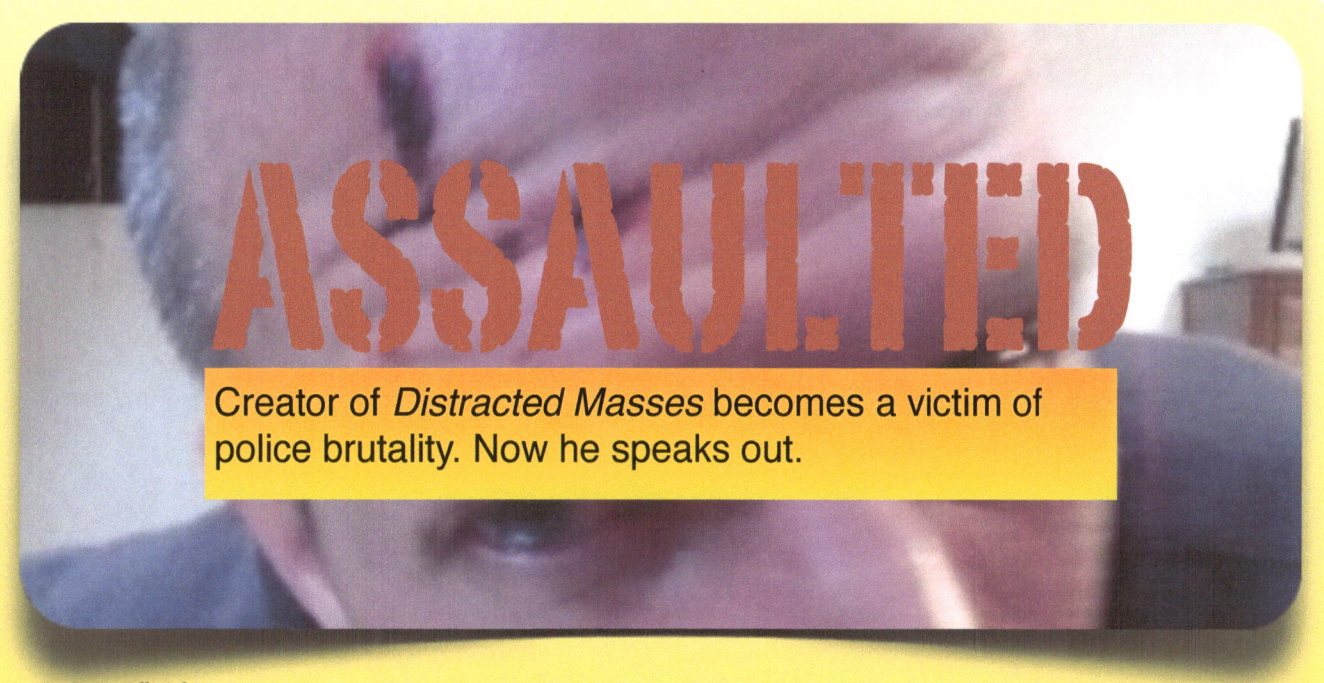

ASSAULTED

Creator of *Distracted Masses* becomes a victim of police brutality. Now he speaks out.

By Scott Albright

What happens when you try to have an innocent night out on the town in Albuquerque? You get assaulted by plainclothes cops, have your phone stolen, get dumped on the side of the road without any medical assistance, and then denied help from local authorities in trying to find the attackers. At least that's what happened to me beginning the night of November 6, 2014 after I went to check out a show at the El Rey Theater. I had just taken the 10:30 p.m. bus from Louisiana & Central to downtown Albuquerque and had two beers at the Anodyne before heading across the street to Rays Smoke Shop and then up to the Stereo Bar. I'd never been to the Stereo Bar before and asked the guy at the door about the show before deciding it was one I wanted to see. He said the show featured an artist named Salva, who played rap & hip hop.

I paid a $10 cover charge, went inside, used the bathroom, ordered one (1) Marble Red 12 oz. beer and went back to talk to the person at the door for about five minutes. I then went into the El Rey through a door that connects to the Stereo Bar and found a table to sit at above the dance floor. About 5-10 minutes later, around 11:45 p.m., three (3) plainclothes police officers showed up and stood behind the table I was sitting at. All three were wearing slacks, button up shirts, and dark colored dress jackets, with one of the officers having a badge along his belt line. The officer with the badge looked

> *"Instead, the next thing I remember, all three officers were approaching me quickly, and then all of sudden I saw a flash of light and BOOM! Everything went black."*

to be about 50-55 years old, with a thick mustache. He was around 5'10" and had the face of a drinker, with a kind of large nose and pinkish-brown complexion. He was a little heavy-set and pudgy around the gut. The badge was yellowish gold and rounded at the top. At the time I didn't recognize the badge and thought maybe they were with the FBI. The officer who was wearing the badge went down and across the dance floor into another section of the building before coming back to the place behind my table where I was still sitting by myself. I remember seeing his profile clearly while he walked across the dance floor and remember him having poor posture, with an almost limp-like walk. After he returned from the other side I became a bit nervous since the three officers were standing so close over my table, but still thought it would be appropriate to ask them what they were doing there.

"Would you like to have a seat?" I remember asking as I looked over my shoulder to get a good look at them. There was no attempt to answer my question. Instead, the next thing I remember, all three cops were approaching me quickly, and then all of sudden I saw a flash of light and BOOM! Everything went black. I completely lost consciousness from what I believe was some kind of stun gun or taser to the head shot by one of the officers. This was at about 11:55 p.m. because the show started at midnight and I had just checked the time on my phone, remembering Salva was about to come on.

Instead of seeing the show, I ended up unconscious for several hours, from 11:55 p.m. until about 5 a.m. when I came to WALKING west on the south side of Lomas, just east of University. When I came to I was very disoriented and unsure of where I was until I saw the 7-Eleven near the intersection of Lomas & University. Almost as soon as I walked into the gas station the clerk asked me what had happened to me, as I was oozing blood out of my forehead. After seeing my wounds in the surveillance video monitor and trying to clean myself with a couple of napkins, I purchased a pack of cigarettes, left the store, and walked up Lomas to Louisiana, after first getting a bite to eat at the McDonalds on San Pedro & Lomas. The sun was just above the mountains when I finally made it home and walked through my front door, at which point I was exhausted, but still able to get on the computer and record what I thought happened to me the previous night. I posted a picture of my head wound on Facebook and Twitter and wrote that the cops had bashed my head, before falling asleep still bloody and disoriented.

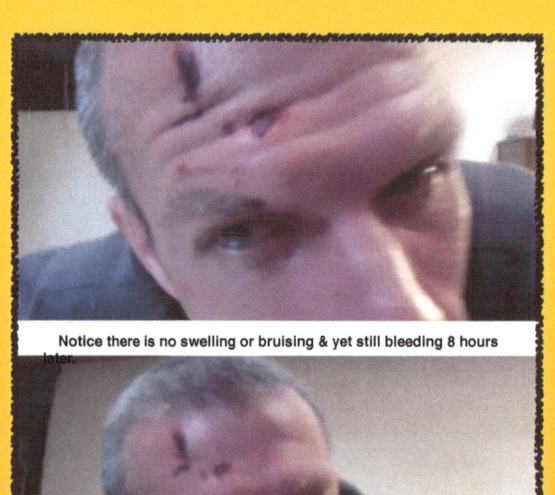
Notice there is no swelling or bruising & yet still bleeding 8 hours later.

After waking up I picked my kids up from school and went to my parents house where I stayed the night to recover from the incident. While at my parents I did everything I could to try and figure out why I was attacked and exactly what happened. I kept remembering the word Imbibe and a face of an older white male with a police badge. I've never been to Imbibe before but discovered it was the location at which an off duty Bernalillo County Sheriff's deputy was beaten for identifying and reporting a person who supposedly had a warrant out for a drunken driving incident. What I found out was that one person who was involved in the assault on the deputy had been arrested while the other person involved got away. I also learned that Imbibe plays similar music to that of the show I was going to see the night I was assaulted and that some of the same DJs and clientele from Imbibe may have also been at the Thursday show downtown. It was my belief, or the best theory I had to go on anyway, was that the cops who assaulted me were looking for the person

involved in the deputy assault who had gotten away and decided to retaliate against me before confirming my identity.

After discussing the incident with my parents I decided it would be wise to file a report with the police, so on 11/09/2014 I filed an online report with the Albuquerque Police Department (APD), describing the incident with as much detail in the limited space available. The next day I received an email from APD saying my incident report was rejected because I reported it as a theft/larceny instead of strong arm robbery, and that I should report the incident by phone. On 11/11/2014 I went back to my parents' house to use the phone to file the report using the phone number provided in the email. My frustration began to build when I couldn't get through to a real person and had to leave a message using the automated voice service instead, leaving my parents phone number for them to get back to me. Soon after I hung up the phone rang. My dad answered and told me the person on the other line was asking for me by name. To my surprise, the person on the other line was calling from the National Police Association! Maybe it was coincidence, but there is no reason the National Police Association should be calling me at my parents house, so I couldn't help but to be taken aback from the call. The person on the other line began telling me he was going to send a decal to put on my car so I could show my support for their organization. I stopped and asked the caller how they got my number and if he was calling to help out with a crime that was perpetrated against me, asking if he was attempting to help a military veteran in need on Veterans Day. He said no and how great of a job the police were doing before I told him I didn't want the decal and got off the phone.

On 11/13/2014 I drove to an APD substation near Louisiana & Zuni where I proceeded to go inside and tell the white, heavy-set, older lady behind what could only be bullet-proof glass that I wanted to file a police report regarding the Nov. 6 assault. The lady behind the glass told me I had to call 242-COPS. I told her my phone was stolen and that was why I was reporting the incident at the substation instead of over the phone. She then pointed to a phone in the substation that I could use. After dialing the 242-COPS number using the substation phone I was finally connected to someone who said a police officer would be out to talk to me. After waiting for about one (1) hour, in which time the lady behind the desk personally helped several people, a young 30ish looking APD officer named Patrick Casias walked out the door inside the substation and asked me if I wanted to sit down. After listening to me tell my story Casias handed me a card with his information and the APD Case Number 140103807 written on the back. He promised to look into the case and that if I had any further information I should contact him. He also advised me to get checked by a doctor to see if I'd been drugged, which I did almost immediately after leaving the substation.

At the VA I was examined by a nice female doctor who seemed genuinely concerned about my well being. She told me that if I had been drugged the type of drugs that could cause the type of memory loss I experienced would already be out of my system. She asked if I wanted a brain scan, but didn't think it was worth it because it would only show hemorrhaging and if that was the case I would've probably been in the hospital a lot sooner. She examined the two marks on my forehead which initially looked like two puncture wounds about a half inch apart, but by the time of the doctor's visit they had pretty much healed and looked like a couple of round scabby dots, partially covered by wrinkles. The doctor agreed that it sounded like I had been tasered and that the memory loss was indeed something that could happen from being shocked in the head with enough voltage. She said my wounds didn't look like I was beaten with a blunt object as most type of blunt trauma to the head that could cause that type of memory loss would have left bruising, swelling, or other identifiable marks which weren't present on me at the time. Not to mention, I've been knocked out in the past after having bashed heads with another person while playing in a soccer game, and not only did I regain consciousness much quicker in the soccer incident, but I also had a much bigger wound that had to be stitched up at the hospital afterward. I've also had fuzzy memory recalls from drinking, but never from that little amount and never to the point where one moment I don't remember anything at all, and the next minute I'm coherent enough to notice that I'm walking down Lomas. Other than agreeing that it was possible I was tasered in the head, the doctor at the VA could not confirm anything and at

best could only write up a report detailing the incident and the after effects I was experiencing, including painful headaches and continued short term memory loss. The doctor recommended I go back and interview as many people as possible at the bar to see what I could learn. Unfortunately my time to do so was quite limited, but I did attempt to contact a number of people online and began my own web search for the face of the cop I still have burned into my memory. I also filed an online police complaint form with what I now know is a completely defunct Civilian Police Oversight Agency (CPOA). After submitting the complaint I eventually received a response from someone who called herself Diane (McDermott, I believe). Although Diane seemed like she was actually going to look into my case, it was more of a false hope than anything close to what we like to call reality.

On 11/22/2014 I recovered my phone by what must have been sheer luck, as I had been trying to trace its whereabouts using Google's Android Device Manager and by calling my number repeatedly with a new, cheaper phone I had to purchase after having my only other one stolen. Although I had no luck with either method, my dad was able to get through to someone who he said sounded like was possibly homeless because of some things he had mumbled, which gave me a tiny bit of hope that I could learn more. In fact, it was a homeless person I saw standing in a parking lot who inspired me to try and call my phone again on the 22nd, and to my surprise there was an answer on the other end. I quickly began negotiating for the phone, offering the person on the other end $45 to give it back. The person with my phone said he wanted more, but I told him it wasn't worth it and that that was all I had. He told me to meet him at the 7-Eleven on Montgomery, just east of Carlisle. After parking on the east side of the gas station I called my phone to let the person on the other end know that I was there. He told me to wait, which I did, but I became bored and walked inside the gas station to get something to drink and a couple of snacks. When I walked outside there was an SUV and a 4-door sedan parked next to my vehicle with about three (3) or four (4) young men inside playing loud gangster rap blasting the lyrics "murder, murder, murder - kill, kill, kill."

"Crap," I thought. "What did I agree to? These guys are going to rob me." I asked one of the guys who had stepped outside of the car if he had my phone, which he said he did. He told me he had bought the phone off some guy on a bike just down the road, and that's all he knew about it. I gave him the cash, thinking he was going to run, but was instead handed my phone before everyone departed and went their separate ways. Unfortunately all the data had been erased and my SD card removed from the phone, but it still worked, as I kept receiving phone calls and text messages asking if I had a 'teener' or if I 'wanted some black.' Needless to say, I turned the phone off and let it run out of minutes. "If the cops look at my phone they'll think I'm a drug dealer," I thought. "This is not good." Despite my trepidation I decided it would be best to update Casias and let him know I'd also identified two persons of interest who could possibly know more about what happened to me the night I was attacked. To add to my frustration, I was told Casias was off for the day and that I could leave a message, which I did. I waited several days before deciding that I should try to contact Casias again, but this time was told by the person answering the phone that the computers were down and that I'd have to leave a message.

No one from the Albuquerque Police Department attempted to contact me through the whole month of December, so I took it upon myself to report the incident to the Attorney General's office in downtown Albuquerque, which was a much easier and less stressful experience than reporting the incident to the police. They actually let me report the incident in person, face-to-face with no wait time! When January rolled around I'd pretty much given up hope. One of the most traumatic experiences in my life wasn't even a blip on the radar screen for these guys. And I've had some pretty traumatic experiences in my life, like being shot at in Iraq and seeing dead burned bodies littered across the highway, but this was something else. I couldn't think about anything else other than the face of the officer and the time that was missing from my life. I was pretty much fed up, but then on Jan. 8, 2015 I received a call at 6:42 p.m. from Diane, the investigator with the CPOA, phone number (505) 924-3727. My hope was revived, as Diane was the only person I had talked to about the incident who I felt might actually be able to do something . . .

"Hello," I answered.

"Can I speak to Scott Albright," the voice said, identifying herself as the Diane who had contacted me earlier regarding my police complaint.

My excitement peaked, but then flattened out not even a split second later. "I'm sorry to tell you" . . . blah, blah, blah. . . No information, no interviews, no video, no nothing. Diane hadn't talked to anyone. Not a single person. She was calling to tell me the CPOA was done and wringing their hands clean of the case. I interrupted before she tried to get off the phone.

"Just for the record," I said. "After reviewing everything that happened and looking into the

matter further I've come to the conclusion that either a taser or stun gun was used on me when I was assaulted."

Diane interjected. "If it was a taser there would be two puncture marks. . ."

"There were," I said. "About a half an inch across, not too far apart."

"Oh, they would be further apart," she retorted. "If it was a stun gun there would be burn marks."

I didn't even wonder if she'd considered that the width of the marks should be smaller since the officers were at such a close range, maybe 2-3 feet away. Of course she hadn't. She had no information. She hadn't talked to anybody or even looked at the picture of the puncture marks on my forehead I had taken almost as soon as I came home the morning of Nov. 7. She told me herself that she had no information, yet she continued to argue with me about my own account of what happened, like she knew more about it than I did.

"Nothing like this has ever happened to me before," I said. "I've never lost my memory like that."

"Tasers don't cause memory lose," Diane lied.

I started shaking at this point. What the hell did this lady know about anything? Why the hell was she calling me? To lie to me? Of course tasers cause memory loss when shot into someone's head.

"I'm sorry, but I've done the research and looked at the medical journals," I told Diane. "Tasers do cause memory loss. There was a cop who was shot in the back of the head with a taser who lost his memory for an hour," I told her.

After reviewing the Canadian Medical Association Journal article I was referring to a second time, I realized the actual time of memory loss the cop suffered was probably much longer. "You and I are not jiving," I told Diane. "Now you are just telling me untruths." I hung up the phone and proceeded to let the shaking out of my body before attempting to vent my frustration elsewhere.

Unfortunately, the most I can really do to vent is write this article so as to inform the citizenry about those criminals who continue to patrol the streets as impostor law enforcement agents, carrying deadly weapons, badges, and an arrogance made deadly by their lack of conscience and compassion. At least I can say that I warned you and that I did my part in speaking out, no matter how much it may hurt me to do so. Perhaps I will never know who attacked me on the night of Nov. 6, but I still want to know what the hell is going on. Were these police who attacked me out to retaliate for the assault that happened at Imbibe the previous month? Did they mistake me for the person who is still on the loose and then only realize it after they assaulted me? And why has it been so hard to file a police report of this incident? Who do I have to go to for someone to investigate? What happened to me is wrong in every way possible and I want some answers. I was the victim of a crime but I can't get the police to help me figure out who these criminals are so they can be prosecuted to the fullest extent of the law. There are criminals wearing badges out assaulting people for no reason whatsoever and there isn't a care in the world. This makes the whole police department culpable. From my perspective they are all criminals and guilty by association. The only thing legitimate about the law is the bullets that back it up. Everything else is a farce, except for perhaps the cold prison cells where they've amassed our young, poor, mentally ill, and addicted into what could only be likened to the overstuffed cow pens where cattle are herded before going off to the slaughter house.

Despite my cynicism, I will continue to pursue this matter to the end, but am doubtful there will ever be any kind of resolution. Just think, I woke up walking down the sidewalk on Lomas and still have no memory of the time between then and when the police approached me. This has never happened to me ever in my life before. What if I had woke up walking down the middle of Lomas and been hit by a car instead? I would be dead. Why wasn't I admitted to the hospital? Why can the police continue to get away with these crimes even after the Department of Justice spent so much time trying to get the police in Albuquerque to conduct themselves in a more professional manner? This is corruption to its core. I am disgusted with the way I was treated and can't stop thinking about it. Unfortunately nobody else has even bothered to think about it at all. Perhaps they will when it happens to them. But by then it will be too late.

UPDATE (3/11/2015)

After recording all that's happened to me since Nov. 6, 2014 in the above article I was finally contacted via certified mail by the city of Albuquerque as a response to my criminal complaint. In sum, I was told APD cannot investigate crimes that were committed against me in the city of Albuquerque if it was by a police officer employed by another agency other than APD. This was signed by the chief of police himself, Mr. Gorden Eden. So the local police cannot help me to solve a crime that was committed against me in their jurisdiction? Is this for real? Who am I supposed to tell my kids who to call if a crime is ever committed against them? Certainly not the local police. They can't help. And this isn't the first time I've learned that, so I won't waste my kids' time teaching them to memorize the 9-1-1 number. If I were a libertarian I'd have them memorize the insurance company's number instead, but unfortunately for us, I'm not wealthy enough to be a libertarian, so they'll just have to memorize my number instead. Having served four years in the Marine Corps I feel confident that I can react to any emergency just as quickly and effectively as the local police, but I'm not a member of any gang, vigilante group, or local militia, so unlike the cops, I don't have backup. As a Marine I was trained to use overwhelming force against those trying to kill or hurt me, but as a civilian I can't do this legally, so my Marine Corps training does me no good at this point. Anyway, I don't believe it would be appropriate to shoot or taser the guy in the head who tasered me, as I believe it would just perpetuate the violence and put my children needlessly in harms way, but still, how in the hell am I supposed to find justice?

I've considered going to the county cops and the state cops and even the DOJ or FBI, but what's the point if I'm going to get more of the same bullshit? Maybe I can contact the U.N. and get the Russians to do something. Perhaps not. What I do know is I will not seek justice through a taser gun of my own. This will not do. Something more is necessary. Something that makes a real change.

UPDATE (3/26/2015)

To date I've received two letters from the City of Albuquerque regarding this case, one signed by APD Chief of Police Gordon Eden. The second letter was sent in reply to an email I sent to Mr. Eden after having received the first letter telling me there was nothing APD could do for me. I had asked Mr. Eden how he knew for a fact that the cops who attacked me were not in his department. Like Diane, Mr. Eden's reply was that they knew this to be true because of the information I had given them. But earlier Diane had totally disagreed with me, telling me my version of what happened was wrong, so I find it hard to believe that they really take my word for what it is - the truth. Despite this, Mr. Eden did give me a good lead by telling me that my description of the badge fits that of the Bernalillo County Sheriff's Office (who I suspected to be involved all along) or possibly the U.S. Marshall's, which I hadn't considered. But after getting online and looking up U.S. Marshall's who served in New Mexico I found a person who matched the description of my attacker to a tee. I won't mention his name because I don't have enough information to pinpoint him as the assailant, but to my surprise the guy I found is an employee at the Kentucky Federal Emergency Management Agency. He had served in the U.S. Army, served as a criminal investigator for the New Mexico Education Department, and was a former U.S. Marshall for many years. I could only find the one picture of him and have no new information. There's no point in telling Mr. Eden what I found, as he was also a U.S. Marshall and may even be friends with the guy. So instead all I can do is pass along what I know to you, the reader of *Distracted Masses*. The wanted poster on the next page uses a digitally rendered version of the guy's picture I found online, because if the police were ever serious enough to do a police sketch this is exactly what I would want it to look like. Like I said, I don't know if the picture I found is of the person who attacked me, but so far this is the closest I've gotten to finding out who did this to me, but it makes my theory about the Sheriff's office being involved seem wrong, so now there are even more unanswered questions and still two other cops whose descriptions I have not been able to match up with any photos.

UPDATE & THOUGHTS (4/26/2015)

As of today, Sunday, April 26, 2015, no new information about the horrendous, despicable, and cowardly assault on me has been passed on from the Albuquerque Police Department or the Attorney General's Office. I have purposely held off on publishing this article on the premise that it could interfere with any investigation into this attack, but it is clear now that there will be no attempt to follow up on this case or for authorities to carry out any other investigations. Again, it is the sad gang-like, mafioso style no-snitch policy in place among U.S. law enforcement agencies which allow episodes like the one I encountered to continue to sabotage and destroy the justice system in this country. No doubt I have some fear about telling my story because of the corrupt and inept system we have to live under, but as an ex-Marine, soccer player, and skater I find it nearly impossible to let fear dictate my life. In the end we all die, after all, and I don't want to die leaving my children thinking the fear these criminals instill in the public is greater than the love, unity, and resilience of the masses who far outnumber these sick individuals who pretend to serve and protect us. Fear is a control mechanism though, and I refuse to be controlled by these assholes who do not have the best interests of me, my family, or my country in mind. If we let fear dominate

WANTED

For the Nov. 6, 2014 assault on the creator of Distracted Masses

Crimes committed include assault with a deadly weapon, aggravated assault, strong arm robbery, theft/larceny, conspiracy, intimidating a witness, tampering with evidence, dereliction of duty, abuse of police powers, corruption, and negligent use of a weapon. If found make a citizens arrest or turn him into your local vigilante group, as he seems to be above the law and immune to the legal consequences of his criminal actions. Be careful. He's armed and dangerous!

Continued: UPDATE & THOUGHTS (4/26/2015)

us than they will continue to kill, maim, torture, steal, rob, cheat, lie, and attempt to destroy our lives. I would rather be dead than live under such fear.

A solution is needed though. Like I said, using the U.S. government training I received to bring about justice will not do in this case. Although I was trained to be a killer and to kill my enemies, I know that such an act would be pointless and would not bring about the necessary changes that would keep my children and grandchildren safe from these criminals. I do think the officers (or FEMA workers) involved in this attack should lose their jobs and go to prison for a very long time, but I know no judge has the balls to actually do their job and lock these thugs behind bars where they belong. I know using the justice system to try and bring about justice would only be a waste of my time. What is needed is something even bigger. Something more earth shattering and systemic changing. What is needed is more than just an alliance of do-gooders trying to make positive change. What is needed is a complete dismantling of the system and a new one put in it's place.

There are already some great ideas about how to go about doing this. The movie Zeitgeist is a perfect example of the alternatives we could use if we only put our minds to it. There is no real reason for us to be exploited by the elite one percent, other than for them to hold on to their positions of power. We have the technology, the capacity, and the means to live without war, police violence, and slave labor. There is no reason for us to be controlled by fear anymore. Our city centers is where we should start. Let's replace the bars, banks, and bail bond businesses in our downtowns that help perpetuate this inept system with health food stores, libraries, urban farms, parks, recreational facilities, and multicultural centers. Let's let the robots do the slave labor while we do the things that build community and create peace. Let's demand that we are valued as human beings and that all of our time on this planet be treated as it is just as important as the net persons'.

Sure, there are bad apples out there who want war and destruction and power over all else, but they are the few and the weak. As the population grows beyond seven billion there will be more Hitlers, and Stalins, and Bushs, but there will also be more Ghandis and Dali Lamas and Mother Teresas. Humans are driven by love much more than by fear, so we know who the true dominant types are - it's just a matter of coming together now. Through our conscious efforts we can bring peace to this planet. We can make the systemic changes that prevent criminals from infiltrating the ranks of the protectors of the peace. We can stop the psychopaths from rising to power, and we can create a system that creates more opportunities and successes. We can rise up and change the world! All we need is to awaken from our slumber, see through the lies, and seek freedom from the mental slavery we've become so accustomed to. All we need is to come together as a species whose desire to survive outweighs the desire to rule and control others. All we have to do is use our brains, our collective consciousness, and our love. Not only can we, but we have to. For our future's sake.

FINAL UPDATE -

After waiting forever to hear back from the Attorney Generals' office, I decided I better call them to see what was going on with my complaint. Perhaps I shouldn't have, as that would've allowed me to hang on to some tiny little bit of hope that someone in our state's leadership might actually give a damn about the people they're supposed to serve. But I guess it's better not to hang on to any type of false hope, as that would be like living in a reality that doesn't exist. To my shock and surprise, the Attorney General's office had sent my complaint back to APD after APD and the CPOA already told me they can't or won't help me. Wow! I told them I'd just file a report with the FBI since I had a solid lead on who I thought had done this to me. Instead, I looked at the FBI online criminal complaint form and thought, "What the fuck? These people aren't going to help me either." So then, a few weeks back I decided to use LinkedIn to contact the person I believe tasered me in the head. The person I contacted responded by writing, "I've never been a cop in Albuquerque! I was an investigator for the state but only in Santa Fe and only for a short time. Your unfortunate mishap was not of my doing. Good luck on your future endeavors." When I tried to reply to tell him he should try to help me find whoever did this to me since he had been an investigator and was in a position to help, we were no longer LinkedIn connections. You'd think he'd want to help since the person who did this to me looked just like him. I mean, I sure would. Wouldn't you?

A Night(mare) Out in Burque

One day there was this young man who wanted to have fun,
He took the bus and rode alone among a family of one,
A community of people, unique individuals, a crowd to talk with,
Heading downtown to meet more great humans,
To watch a show, to dance, to groove, and feel the rhythm of life.

He gets to this place and starts to mingle,
All over he meets nothing but wonderful people,
He buys a drink, takes a seat, and waits for the show to begin.
Nothing wrong with this picture, no vice at all, wouldn't even call it a sin.

But to his dismay, three cops appear.
But this young man, he still has no fear.
He's done nothing wrong and wonders why the cops are here.
He offers them a seat, and they start to come near.

But then BOOM! Lights out - the cops did something to this man.
He awakes far away walking down the street.
Wondering where his memory is, why the sun is about to rise.
He steps into a 7-11 and buys a pack of smokes.

The clerk asks, "What happened to your head?"
The man realizes he's bleeding, but thinks he must already be dead.
Shaped like a bullet wound - blood oozing out.
He must've been shot and woke up in hell.

But no, wait! Don't jump to conclusions.
This man is still alive and actually feeling quite well.
Bits of memory are recalled as he walks his ass home.
Soon he realizes he was only assaulted, it wasn't a bullet to the dome.

Now he feels even better, he remembers a face.
An older white male wearing a badge.
He remembers the words Imbibe, but he doesn't know of the place.
Looks on the computer and starts to figure stuff out.

These snarky pigs made a mistake!
They were there to retaliate.
But they got the wrong guy, it wasn't the one they were after.
   Now this man has to control his laughter.

How could they be so stupid,
To just misplace they're mistaken identity?
Probably wondering how he can even be remembering.
With that blow to the head or the memory loss drugs he was given.

Now this man doesn't mind the beating, the cops are forgiven.
But he'd like his phone back.
Wondering if any one can help him to retrace and track.
Could even matchup spacetimes' of the phone and those of potential attackers.

To do this he even thought of recruiting some hackers.
But instead he's turned back to the law,
Asking wouldn't the cops prosecute if they were also beaten?
Or would they just retaliate because the law's just not speaking?

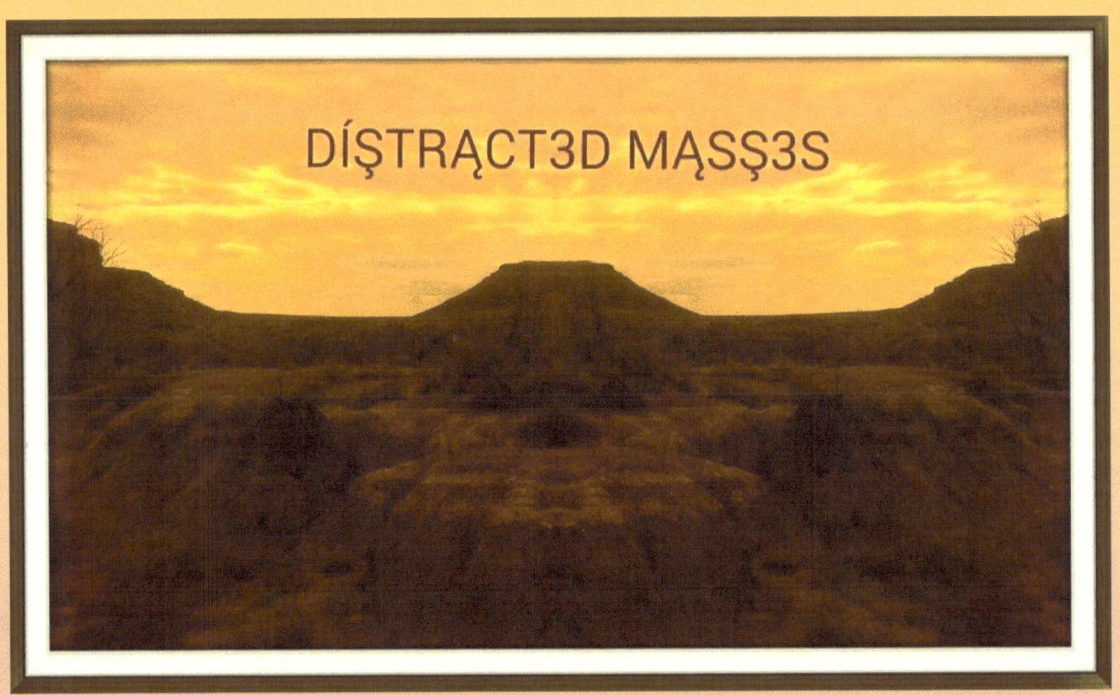

Support The Mission

Visit the Zazzle product pages throughout this issue to shop for cool stuff, purchase a print copy of this mag, buy one of the books listed within, visit the Distracted Masses website and click on ads, donate, or submit an inquiry about how to place an ad in the next issue.

Buy some art at zazzle.com/scajax & zazzle.com/distractedmasses or make a PayPal donation to distractedmasses1@gmail.com.

Distracted Masses is building an audience. Here's the stats rundown as of July 12, 2015: Scribd, 531 views of Distracted Masses Vol. 1 Issue 1, 426 views for Issue 2. Facebook: 919 likes. Twitter: 456 Followers. Tsu: 1K Friends, 180 Followers. Weebly site on Google Analytics, 06/11/15 - 07/11/15: 1,128 page views, 845 users. ISSUU: 936 Impressions. Distracted Masses is also on YouTube, Newsvine, and other prime advertising spaces. Contact distractedmasses1@gmail.com to discuss rates and help grow an even larger audience. Thanks for your support!

Ponderings & Poems

On Life & Death
by Alexander Albright
March 9 at 11:34am ·
"Everything that we know depends on time. There is no such thing as forever for us living things. Because everything that is alive will die. It will, and there is no escaping it. The fact that we cannot comprehend a forever is what makes death such a scary thing for us. We cannot accept the fact that one day we will cease to exist from this world. And yet no one really knows what happens to you when you die, I mean sure, we all know what happens to your body. It decomposes, and over thousands of years it will become a nothing, just providing minerals for the earth. ~ But what happens to YOU?" - Life and Death by Alexander Albright part 1/3
#deep #staytuned #lifeanddeath

March 10 at 1:14pm ·
"Sure we die, but our energy, our spirit, will not die. We will go on forever. Why? Because energy does not go away, it just changes form. When an energy is created, it won't go away but simply just change and transfer. In this sense, ~ ~ ~ we are forever. We will always exist." - Life and Death by Alexander Albright 2/3
#deep #staytuned #lifeanddeath

March 11 at 9:49am ·
"What happens and what it is like to be in your 'spirit form' (what happens when you die) is just something that cannot be experienced or learned about until the time comes. It can't. It is the one thing in life that no man can find an answer to. That is what makes it so frustrating, and yet scary for us. The fear of the unknown. In order to get past this we must cope with a more simple answer to it all, and that is 'one day we will

find out.' And until then I believe that our purpose is to find the joys in life, and to be the the best person that we can be. ~ Is that not the whole point?
- *Life and Death by Alexander Albright part 3/3*
#deep #finshed #lifeanddeath

Reality?
by Rick Albright
The pine filtered sun
ascends from the east
and places a beam

on a speckled
woodpecker
drilling into a tree.

My senses
perceive
the marvels all
around
while my brain
decides
what wonderful
things
it expects to
perceive.

Quantum theory
provides
an explanation of
sorts,
It is all subatomic
particles
spinning and whirling
magnetic flux.

influenced by

It doesn't really matter
what it is or how
it came to be
for beauty doesn't care
about it's beginning.

Excerpt from Wyrms by Orson Scott Card
And we believe this story, of how everything is causally connected, without questioning it......And this is the world we live in, this pattern of events, that cause each other. It becomes the framework by which we remember everything. But some things come along that don't fit.

The Story
by Rick Albright

Overtime it grows to fit the needs of the teller. Our brains select and edit memories dependent upon our need.

The young marine heading home from a terrible war tells a story in the barracks of Danang- So we opened fire we moved forward as we carried him on our backs while the RVN sniped at us while we found cover- Was pretty close to the facts and served the need to relieve the past.

Later in life he says- So I opened fire and moved to the front where I picked him up and carried him away while they shot at my back.
-the pronoun has changed the others are gone, his needs have changed.

He is no longer relieving the past he is reliving the past for others to see him in the light he he needs today.

Complete Poems, 1924
by Emily Dickinson

THE BRAIN is wider than the sky,
For, put them side by side,
The one the other will include
With ease, and you beside.
The brain is deeper than the sea,
For, hold them, blue to blue,
The one the other will absorb,
As sponges, buckets do.

The brain is just the weight of God,
For, lift them, pound for pound,
And they will differ, if they do,
As syllable from sound.

Never Forget
by Anonymous

We never forgive, We never forget.

Why turn the other cheek? When they don't give a shit?

Dry

Dried out like a raisin.
Radioactive
Metal
Heavy
Depletion
Decay

Sun baked,
Cooked to a prime.
Test site.
Trinity.
Dried out.

Sunflower Water Bottle

David J. Hand's *The Improbability Principle*

Review By Scott Albright

Probabilistically I'm sitting here writing these words, but just as probabilistically I'm also sitting somewhere else writing these same words, or even some other words for that matter. I can be in the same place at the same time, but can't I also be present in two times at the same place? Or present at two times in two places? In extra dimensions I suppose it's possible, but in David Hand's *The Improbability Principle* we're not going beyond the human experience, which in most cases is only in four dimensions. This is a good thing for understanding everyday life, as this book brings home some basic principles we need as humans to sharpen our survival skills.

First off, there is the law of large numbers. Let's begin with that. When it comes to the likelihood of us dying there is a good chance it will come from an automobile accident, heart disease, cancer, a natural disaster, or some form of violence. Now there are seven billion people on this planet. If the majority of us die from the above mentioned things than it should be quite obvious that all we have to do to improve our chances of living longer is to avoid the things that cause those forms of death. So that means no driving or being near cars, staying away from junk food, cigarettes, alcohol, and drugs, avoiding areas prone to natural disasters, and staying away from violent environments. In doing so we can avoid being another one of the large numbers statistically proven to meet the inevitability of death because of the above mentioned things. That brings us to what I think is the second most important point of *The Improbability Principle* - the probability lever.

Manipulating the lever can change the outcome of events by means of the butterfly effect. Small changes can have large effects over long durations. When mixed with the

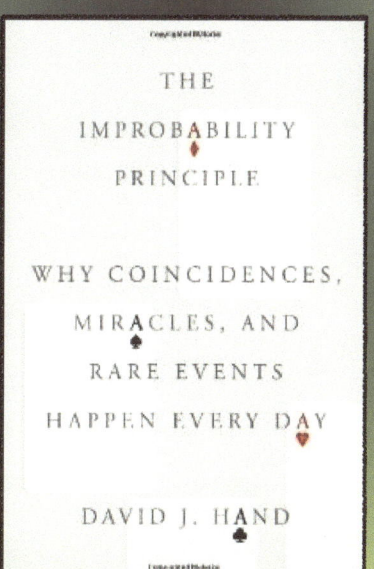

law of large numbers and the law of selection, the probability

lever can also create dramatic effects in short periods of time. Measuring the outcomes may not be so easy however, as the outcomes may never be as predictable as one thinks they will be, no matter how refined their alterations to the probability lever are. Beyond trying to control the probability lever for survival purposes, one could also use it to try to change the outcome of a political election, as I had considered while writing a piece on the Bernalillo County Sheriff's race during the midterm elections.

After writing the article, in which I thought it was nearly impossible not to use the law of selection to provide information which made one candidate seem more favorable than the other, I asked my friend to read the article and to tell me who he would vote for based on what I had written. To my surprise he picked the exact opposite person whom I thought he would. Clearly my understanding of the law of selection was wrong. Or was it? As it turned out, the person I thought was going to win based on the information I had available to me, did indeed win the election. But how am I to know if my consideration of the probability lever had any effect on the final outcome? The answer is, I won't. And that's the essence of Hand's book.

You can choose to play the lotto or you can't. If you don't you won't win. If you do, you might, but if you are the lucky one who's to know which factors played the most influence on the final outcome? How much was luck and how much had to do with the law of selection or the manipulation of the probability lever? If you buy all the lotto tickets and win than you know the law of selection and the law of large numbers were in effect, but if you buy one and win it seems like chance is more in play than any type of pre-selected outcome. Card counters can create deterministic futures, but only to a certain degree. And that's the lesson learned from *The improbability Principle* - chance has to rule, as there really is only so much time to try and manipulate the future while in the present. In a quantum holographic extra-dimensional multi-verse one has the time to consider all the possibilities before making a move, but for us Earth-bound four-dimensional human creations time is limited and random chance may always be our only best bet.

So if you want to win the lotto, don't think reading this book will improve your chances dramatically. But if you want to understand why things happen happen the way they do even when they seem they shouldn't, than this book is definitely for you. In sum, the author explains that if the first explanation seems like it is the most improbable, than you should definitely go with something else that is more probable. In doing so we can avoid making some of the most basic mistakes while trying to decipher the different realities constantly being presented to us. I personally couldn't put this book down and recommend it to anyone and everyone who hated math because they didn't understand how it related to everyday life. This book will put an entirely new perspective on that attitude. Read it or don't. The final outcome is your choice.

Steel Nerves
7 1/4" Skate Deck

zazzle.com/distractedmasses

Review: Andy Greenberg's *This Machine Kills Secrets*

By Scott Albright

Before picking up Andy Greenberg's *This Machine Kills Secrets,* I was reading a book about the Rothschilds, which really captured my attention when I came to the part about the control of financial information by Rothschild employed mail couriers in early 20th century Europe. But when I found Greenberg's book the Rothschilds suddenly became very uninteresting. Here was a book about modern day cyber villains working to take information control out of the hands of people like the Rothschilds. Fascinated, I put down the history book and read Greenberg's instead.

Upon finishing the book my first thoughts were that they were all stupid - the hackers, the cyber security experts, the government agents - all of them. What a waste of resources trying to find the most efficient way to be anonymous. I mean, I can understand how the U.S. government could use DARPA created anonymous tools like Tor to hide the identities of secret agents and foreign dissidents. I get how Julian Assange wanted to keep the identity of whistleblowers hidden, and I even get the cynicism and dark humor of some of the cyber libertarian anarchists, but still, why go through all the trouble developing anonymity software if all it's going to do is aid the enemy or opposing party? Do the oppressive and violent forces of 21st century society really scare so many people into hiding? Wouldn't making oneself known be more effective in obtaining the sought after goals than hiding behind a mask? In Gene Sharp's *How Nonviolent Struggle Works,* he argues that anonymity can cause setbacks in a group's cause, and is largely unnecessary in countries where freedom of speech and assembly are protected by law. Sure, he argues that anonymity can be useful in some cases, but still disadvantageous when it's not needed. So the question is, is anonymity really a necessary condition for transparency?

The hackers discussed in *This Machine Kills Secrets* certainly think so. It seems counterproductive though to spend so much time trying to expose and open up information to the world by ensuring those who reveal it stay hidden. Not just counterproductive, but hypocritical in a sense. Yeah, I get it. Information is dangerous *and* valuable. That's why the Rothschilds wanted to control it. That's why Chelsea Manning is in prison and why General Petreus is under federal investigation for sharing classified information to his mistress. It's why Snowden is considered a traitor and why so many of the corporate elite have investments in or managing control over large sections of the global mainstream media. Certain information can cause chain reactions, mass panic, or confusion. The more hidden the information, the more valuable it seems to be.

And we need anonymity and secrecy. Without it bank accounts would be open for the plundering, voters would be unable to make honest selections at the ballot box, and nuclear codes could be hacked by anyone with a cell phone. I don't think the cypherpunks and crypto anarchists will have an easy time convincing the world their goals are ultimately beneficial to society at large though. There is some moral reasoning behind providing an outlet for whistleblowers to release information while keeping their identity hidden, but Greenberg's book tells me many of the hacktivists out there aren't trying to create a secure witness protection program. Instead, it seems, they're trying to expose the world's secrets without considering the true ramifications of their actions. They will never know if their actions have done more good than bad because we can't undo history, but my fear is that now people will become more suspect if they try to be anonymous than if they volunteer their information to government and corporate entities. Wasn't one of the reasons bin Laden supposedly found because the house he was at didn't have the internet, while all the others around him did?

Tor, as many of the hacktivists from Greenberg's book will say, is not secure. I believe cyber criminals using Tor will be caught because there are people out there specifically looking for crooks trying to stay hidden. As the the number of computer users grows with the population it will be harder and harder to keep up with all the hidden networks, secret VPNs, encrypted P2P data exchange sites, and so forth, but still, I think the fight for information control is too great and too valuable for governments to just give away state secrets or corporations their newest patents.

Perhaps quantum computers will change the tide in the information war, but even then there will be other quantum computers capable of defending against the opposing quantum hackers. The cryptographic game of cat and mouse will probably go on throughout many lifetimes, but no matter what, there will certainly be an evolutionary change in how we manage and access information in the future. In fact, there have been many of those changes already.

No matter the changes or the popular sentiment for open information and transparency, people like the Rothschilds will still do everything they can to try and control it. People like the Murdochs and the Turners. The Poroshenkos and the Rotenbergs. Their information is their power. It is their wealth. But because there is already so much information in the world it seems like it is becoming less valuable, yet it actually provides more, as it is the barrage of useless and often disinformative content that helps to provide cover for the more valuable, hidden information. In other words, transparency is a tool of anonymity used to secure the power and wealth of the elite, create false trust, and to continue keeping the masses eternally distracted.

Discerning Alien Disinformation
By Montalk

Review by Scott Albright

Aliens aren't real are they? Oh sure, it doesn't matter that we're just one species on a tiny planet living in a short fragment of time in a giant universe that knows no beginning or end. It doesn't matter that the Kepler telescope is finding more and more planets similar to Earth everyday, or that many of those planets have been around for a lot longer than ours. No, humans have to be the only intelligent species in this infinitesimal universe teeming with galaxies, stars, and planets we haven't even discovered yet. No, we have to be alone.

Or are we? The very informational pamphlet-like book I found online titled *Discerning Alien Disinformation* claims there are several alien species right here on Earth, controlling our daily lives! The author, simply named Montalk, claims there are multidimensional sentient beings and other intelligent creatures from different parts of the universe which use various means of human manipulation for their own benefit. Montalk says these creatures can plant disinformation and manipulate major life changing events on different parts of our historical timeline, altering all of human societies' future pathways. They are creatures that can alter the course of history, and our collective memory of it, through multi-temporal disinformation campaigns.

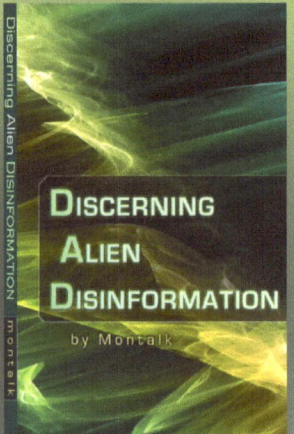

Now if you don't believe in aliens this may all seem a bit outrageous, but if you're one of those whose thought about it and decided that aliens could be sophisticated enough to travel multidimensionally, than this might already sound pretty familiar. I personally don't have any evidence to prove or disprove that aliens exist, but I do think that if an alien species were to visit Earth they would have to have some pretty sophisticated technology to do so, and could have many reasons to keep their existence hidden from humans. For some reason I doubt any of those reasons would be because of any type of a threat we pose to them. Think about it. If humans are still killing each other in wars over resources, what makes you think a sophisticated alien species doesn't also need resources to feed their families and keep them alive and out populating the universe? What makes you think they wouldn't view Earth as a giant farm or breeding ground for their resource needs? Like Nina Paley & Porno for Pyros sing, "We'll make great pets."

But the alien argument aside, Montalk makes some very interesting points about disinformation. Besides the time traveling multidimensional telepathic aspect to it, much of what Montalk claims aliens are doing, humans are also doing. It's no secret corporations use subliminal messages and behavior modification audio/visual systems to influence worker and customer actions. There's no doubt that Big Media is filled with agents and conspirators planting lies and half-truths for political or financial gain. It's obvious that humans too have tried to rewrite or alter the course of history through highly synthetic processes.

And that's what makes *Discerning Alien Disinformation* such a good read. Because learning about the tactics, strategies, and goals of the disinformation campaigns Montalk discusses will help you to also understand them on a human to human level. There are people who get paid to disseminate disinformation, yet honest reporters seeking to find the truth have to write for free or next to nothing. It's not about objectivity or information selection either. It's about honesty. But as the saying goes, "The first casualty of war is the truth." And at war we are. Constantly and always, but will those wars extend into outer space? Are aliens somehow a part of all this killing and warfare? Will humans have to fight them in some intergalactic war?

Montalk doesn't say yes or no to that, but the book does have different scenarios for dealing with future outcomes after the disclosure of alien existence occurs. Some guidance is also provided on how to know if an alien is positive or negative, as there are both types and it's not always easy to tell, so Montalk says.

Perhaps my favorite part of the book is the description of the positive aliens. Montalk says higher positive aliens "possess love, empathy, and understanding" and they fuse their "higher emotional and higher intellectual" sides "into harmony and balance." Montalk says we will notice that these aliens have "a strong degree of sentience, lucidity, earnestness, intelligence, and wisdom," and that "their verbal and telepathic communications resound with truth." In their communication Montalk says that "they are good at encrypting deeper truths into what they say, truths that reveal themselves only in hindsight when one has finally matured enough to understand them, yet they are equally talented at packing much meaning into few words and saying the right things at the right time." Further on Montalk writes, "Positve beings, particularly the middle ambiterrestrials and above, never appeal to your baser instincts or use ego hooks to coerce cooperation. Instead they speak more to your higher reasoning and intuition."

Hopefully these higher positive aliens will have a greater impact on our planet than the negative one's, but if you don't believe in aliens I guess it doesn't matter one way or the other. Unfortunately, disinformation isn't just an alien phenomenon, so it would be wise to read up on all of the potential future possibilities in the dark art, just in case.

Severed Ties

Your cruising on empty,
It's senseless,
To retrace the surveillance,
They were tailin',
Now it's just smooth sailin'.
So why worry,
Paranoia will get you nowhere.

Or are they there?
Could be,
Phone's tapped.
I'm trapped.
Can't get out tonight, or ever.
Sever,
All ties.
Outside communication lines.

Tor for the bored,
It's where the CIA do their crimes.
And where I jump space-time.
I'm fine.
How are you?
Why the funny look?
Are you some type of crook?
A paid thug?
A human bug?
What the?

I can't get out tonight, or ever.
Sever.
All communication lines to the outside.
I've lied.
But you took pages out of the history books.
And gave it all over to your paid crony crooks.
Take one last look.

Cuz I've severed all communication lines.
To the outside.
I've tried.
But you don't listen even though your listening.
So I'm done.
I don't care about your prying eyes.
This is more than just a war cry!

D1STRACT3D MASS3S

Only $24.95

Light Emission Tee

Eyeball Speaker

FUTURE WORLD: A HOLOGRAPHIC REALITY?

Holography & Augmented Reality

As the world's population grows there will be more types of everybody, everywhere, including inventors. It boggles my mind to try and think of the different technologies those inventors will create just in my lifetime, let alone three generations from now. Just in the area of holographic imaging there are tons of ideas coming to fruition already. According to Zebra Imaging's website, "holograms provide a statistically significant performance improvement over traditional textbook materials in understanding anatomy, especially in regards to spatial relationships." If this is true than what about astronomy? Could holograms help to provide us with a better understanding of the cosmos? Can it advance our thinking as a species and help us to understand a higher dimensional universe by bringing the flatlanders to us?

To me it's like going from thinking the world's flat, to understanding that it's round. A whole new set of opportunities await that could change our reality over night. Learning in 2D is one thing, but being able to interact with your learning tools in three dimensions is quite another.

Imagine having a hologram of the Milky Way where you could zoom in and out of any part of the galaxy as it spins around its core. Imagine being able to look at the entire galaxy from any perspective, and any size or scale as it moves in synchronous real time. You could reverse directions and go into the past, or speed off into the future. You could plot and map planets with potential intelligent lifeforms and find the shortest routes to those planets. Or better yet, we could use holography to project our own three dimensional space into fourth and fifth dimensional spaces in order to learn how to get to those planets even faster. Maybe even stranger than that is the possibilities of holography and augmented reality technologies used together.

Combining the two technologies provides an endless set of potentials for practical and not-so-practical uses. Imagine watching a holographic movie where interactive hidden pop ups come out for you to change the outcome of the movie. Or how about a video game where players have to use augmented reality to find secret codes as they control holographic characters on table tops, parking lots, and playgrounds. One could enter an entire virtual world of secret mazes and underground oasis. The line between real and fantasy could be totally blurred out of existence once and for all. People could live out their wildest fantasies in this augmented virtual world of holography, augmented reality, gaming, and social media. Perhaps the internet is bizarre enough for you already, but if techno fantasy gamers ever get their hands on holographic augmented reality technologies, society is going to be in for quite a unique show indeed.

About Distracted Masses

Distracted Masses is the creation of Scott Albright, an independent researcher and freelance journalist living in Albuquerque, New Mexico. Scott's goal is to create a fun, unique, and academic magazine for audiences of all ages and walks of life. Future issues are dependent on the time and resources available to contributors, so if you like what you see make a donation or contribute to the cause.

Future issues will focus on topics such as time travel, extraterrestrials, disinformation, futurism, social psychology, philosophy, science fiction, agenda setting, and media analysis. Send queries, articles, poetry, or other contributions to distractedmasses1@gmail.com.

Crawling Ant Productions

Distracted Magazine is an experimental publication created by Crawling Ant Productions, a web design, presentation, and research business solution for individuals and small organizations. To learn more about products and services provided by Crawling Ant Productions visit crawlingantproductions.weebly.com sometime in the near future. The Crawling Ant Productions website is currently under construction but should be up at some ambiguous later date . . .

Donate

Help to keep the ideas flowing by donating to *Distracted Masses* on PayPal. Send PayPal funds to distractedmasses1@gmail.com.

Thanks:

To all borrowed, third-party content creators. *Distracted Masse* does its best to cite all sources. If we forget to cite you, tell us and we'll get you in next issue.

Want to Contribute?

Distracted Masses wants to hear your story. Send poems, articles, insight, letters to the editor, videos, photos, and anything else to distractedmasses1@gmail.com.

Albuquerque, New Mexico
Crawling Ant Productions @ 2015

Contributors

- Alex Albright
- LaWanda Albright
- Rick Albright

www.ingramcontent.com/pod-product-compliance
Lightning Source LLC
Chambersburg PA
CBHW050841180526
45159CB00004B/1990